Animals
BIG and SMALL

Do you look like someone in your family?
Children often resemble their parents.

Some young animals resemble their
parents and some do not. You can take
a trip to the zoo to see animals and
their young.

The zoo has many animals. If you follow this map, which animal would you see first?

Look at the chart on page 5. How quickly can you walk from the elephants to the monkeys?

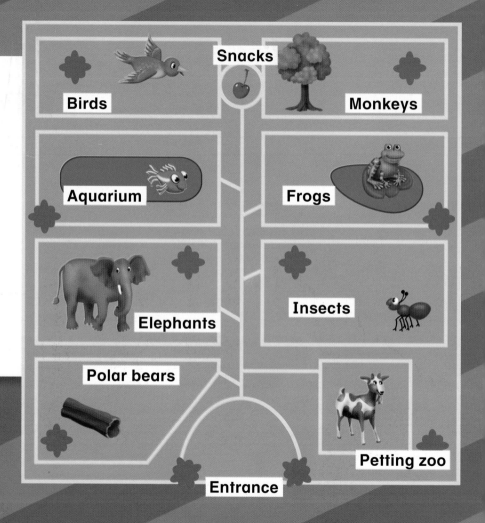

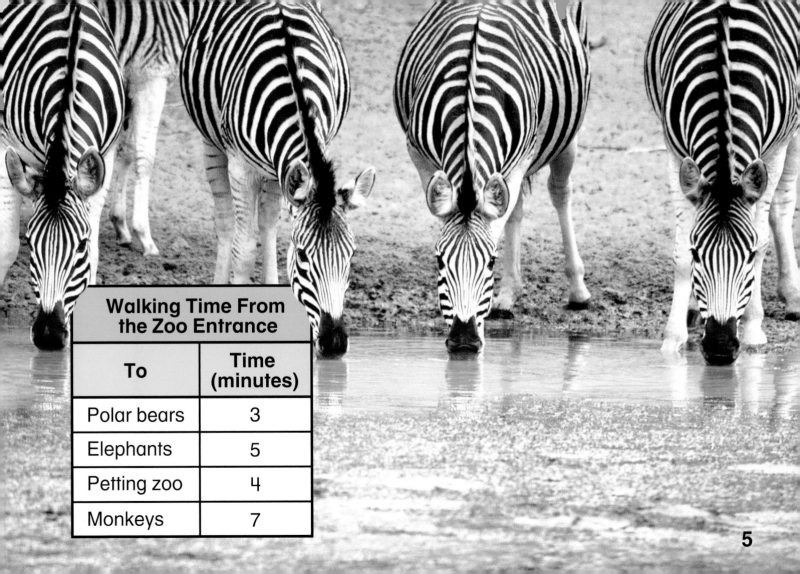

Walking Time From the Zoo Entrance	
To	**Time (minutes)**
Polar bears	3
Elephants	5
Petting zoo	4
Monkeys	7

Baby polar bears resemble their parents. Baby polar bears are about 30 centimeters long. They are nearly ten times longer when they become adults!

Length of Polar Bears	
Polar Bear	**Length (centimeters)**
Baby	30
Adult female	250
Adult male	300

A baby elephant has tusks like its parents. They are called milk teeth. Milk teeth are about 5 centimeters long and fall out after one year. An adult elephant's tusks grow about 15 centimeters each year.

The baby is about 91 centimeters tall.

You can pet goats and sheep at some zoos.
How are these animals alike? different?

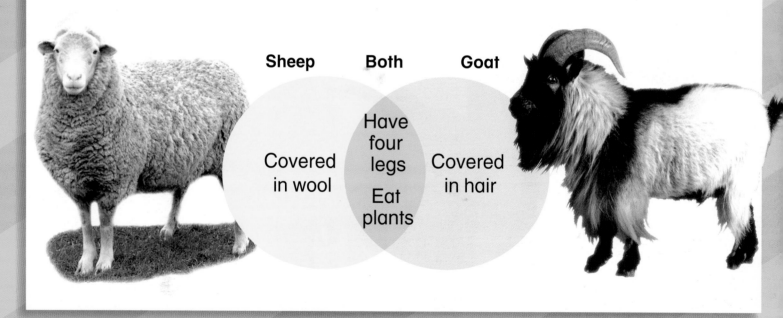

Sheep   **Both**   **Goat**

Covered in wool

Have four legs

Eat plants

Covered in hair

This animal is a baby.
The wool shows that it is
a baby sheep, or a *lamb*.

Many zoos have fish that live in tanks. Seahorses are a type of fish. Some are as small as one centimeter long. Larger seahorses can grow to be about 30 centimeters (1 foot) long.

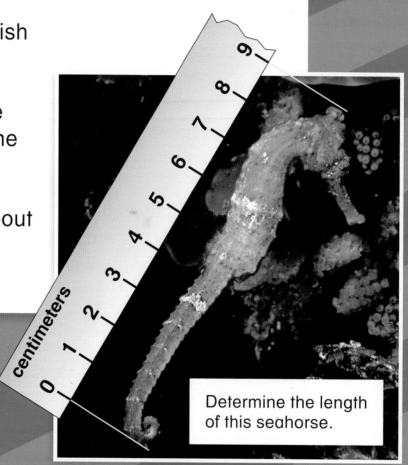

Determine the length of this seahorse.

South American forests are home to the blue poison frog. This frog is small in size, but it is very poisonous.

Most blue poison frogs grow to about 1 inch long. Is the blue poison frog longer or shorter than the crayon?

Which objects are about the size of a blue poison frog?

Compare the adult lions and their cubs. Like people, animals come in all shapes and sizes!